Nature School

LET'S LOOK FOR BIRDS!

By Seth Lynch

Gareth Stevens PUBLISHING

Please visit our website, www.garethstevens.com. For a free color catalog of all our high-quality books, call toll free 1-800-542-2595 or fax 1-877-542-2596.

Library of Congress Cataloging-in-Publication Data

Names: Lynch, Seth, author.
Title: Let's look for birds! / Seth Lynch.
Description: Buffalo, New York : Gareth Stevens Publishing, [2024] | Series: Nature school | Includes index. | Audience: Grades K-1
Identifiers: LCCN 2022045137 (print) | LCCN 2022045138 (ebook) | ISBN 9781538286203 (library binding) | ISBN 9781538286197 (paperback) | ISBN 9781538286210 (ebook)
Subjects: LCSH: Birds–Juvenile literature.
Classification: LCC QL676.2 .L965 2024 (print) | LCC QL676.2 (ebook) | DDC 598–dc23/eng/20221012
LC record available at https://lccn.loc.gov/2022045137
LC ebook record available at https://lccn.loc.gov/2022045138

First Edition

Published in 2024 by
Gareth Stevens Publishing
2544 Clinton St,
Buffalo, NY 14224

Editor: Kristen Nelson
Designer: Andrea Davison-Bartolotta

Photo credits: Cover Bonnie Taylor Barry/Shutterstock.com; p. 1 Mihai_Andritoiu/Shutterstock.com; p. 5 ZoneCreative/Shutterstock.com; pp. 7 (top left), 24 (feathers) PatP66/Shutterstock.com; p. 7 (top right) K Quinn Ferris/Shutterstock.com; p. 7 (bottom left) Joel Trick/Shutterstock.com; p. 7 (bottom right) Steve Byland/Shutterstock.com; p. 9 Valdecasas/Shutterstock.com; pp. 11, 24 (beak) Alexandr Dobysh/Shutterstock.com; p. 13 Don Mammoser/Shutterstock.com; p. 15 Locomotive74/Shutterstock.com; p. 17 Henk Bogaard/Shutterstock.com; p. 19 meunierd/Shutterstock.com; p. 21 Photicus/Shutterstock.com; p. 23 Lin Emma/Shutterstock.com.

Printed in the United States of America

CPSIA compliance information: Batch #CS24GS: For further information contact Gareth Stevens, New York, New York at 1-800-542-2595.

Contents

We hear a tweet.
We hear a chirp.
Let's look for birds!

Birds have feathers.
Some are colorful.
Some blend in.

Barn owls are brown
and white.
They live in trees.
They hunt at night.

Birds have beaks. These are used to catch food.

Hawks have a hooked beak.
They eat rabbits and bugs.
They eat snakes and frogs.

Birds have wings.
Most can fly.

We see a hummingbird.
Its wings move fast.
They hum!

Birds lay eggs.

We see a robin's nest.
It has blue eggs in it.

Penguins are birds too. We see them at the aquarium!

Words to Know

beak

feathers

wings

Index